# The Ultimate Guide to Making Money Online

## Legitimate and Authentic Opportunities for Students, Women, Housewives, and Remote Workers

BY

ZAHEER AHMED SHAIK

B. PHARM., M. PHARM., & C.M.S.-ED

# ABOUT THE AUTHOR

## Zaheer Ahmed Shaik

### *Certified General Health & Business Consultant*

**Zaheer Ahmed Shaik** is a visionary business consultant, accomplished professional, and dedicated mentor with a proven track record of helping individuals and businesses thrive in the ever-evolving digital landscape. With a robust academic foundation that includes a Bachelor's and Master's in Pharmacy and a certification in Community Medical Services and Essential Drugs (CMS-ED), Zaheer blends analytical rigor with creative problem-solving to empower his clients.

## A Multidimensional Professional

Zaheer's expertise extends far beyond his medical background. As a Certified General Health and Business Consultant, he has successfully navigated diverse fields, including:

- Web design and development

- Business growth strategies

- Search engine optimization (SEO)

- Personal and professional skill development

This multidimensional skill set enables Zaheer to provide tailored solutions, bridging the gap between technical know-how and business acumen.

**Champion of Digital Opportunities**

Recognizing the transformative power of the internet, Zaheer has dedicated much of his career to exploring and harnessing the vast potential of online income opportunities. His passion for empowering others has led him to specialize in guiding students, women, housewives, and remote workers toward financial independence. With this book, *The Ultimate Guide to Making Money Online: Legitimate and Authentic Opportunities for Students, Women, Housewives, and Remote Workers*, Zaheer aims to demystify the digital earning landscape and provide actionable strategies for success.

**Why Zaheer Wrote This Book**

Zaheer understands the unique challenges faced by individuals seeking flexible, home-based income opportunities. Whether it's a student looking to fund their education, a housewife balancing family and work, or a professional seeking an additional income stream, Zaheer believes there is room for everyone to succeed online. This book is his blueprint for navigating the digital economy with authenticity, resilience, and purpose.

**A Trusted Mentor and Consultant**

Zaheer's practical, hands-on approach has earned him a reputation as a trusted advisor among his clients. Known for his ability to break down complex topics into easy-to-understand strategies, Zaheer has transformed countless lives through his guidance, workshops, and personalized consulting sessions.

**Beyond the Professional Sphere**

Zaheer is not just an expert but also a lifelong learner and advocate for personal growth. His hobbies include staying updated with the latest technological trends, exploring innovative business models, and mentoring aspiring entrepreneurs.

**Zaheer's Message to Readers**

"Success is not reserved for a select few; it's available to anyone willing to learn, adapt, and put in the effort. The online world is full of opportunities, and this book is my effort to help you find your unique path to success. Remember, the key to thriving isn't just knowing what to do—it's doing it with integrity, consistency, and passion."

Zaheer Ahmed Shaik's journey, insights, and passion for empowerment make him the ideal guide for anyone seeking to unlock the potential of online income opportunities. With this book, he invites you to take the first step toward a future filled with financial freedom and personal fulfilment.

## CONNECT WITH AUTHOR

- **Website:** Zaheer Ahmed Shaik – https://zaheerahmedshaik.online/
- **Social Media:**

  LinkedIn – https://www.linkedin.com/in/zaheerahmedshaik

  Twitter – https://twitter.com/zaskdp

  Facebook – https://facebook.com/iamzaheerahmedshaik

  Instagram – https://instagram.com/zaheer.ahmed.shaik

  Pinterest – https://www.pinterest.com/zaheerahmedshaikonline/

  Tumblr – https://www.tumblr.com/zaheer-ahmed-shaik

  Telegram – https://t.me/zaheerahmedshaik

  YouTube – https://www.youtube.com/@ZaheerAhmedShaik-Official

- **Official E-Mail:** info@zaheerahmedshaik.online
- **Live Chat:** https://tawk.to/zas786

# PREFACE

The internet has revolutionized the way we live, learn, and earn. In today's world, the opportunities to generate a legitimate and sustainable income online are more abundant than ever before. Whether you are a student looking to support your education, a woman seeking financial independence, a housewife managing family responsibilities, or a professional aspiring to work from home, the online economy offers something for everyone.

As someone who has spent years guiding individuals and businesses to thrive in competitive landscapes, I've seen firsthand how the right strategies can transform lives. Yet, amidst the promise of online income lies a maze of misinformation, scams, and half-baked solutions that often deter people from realizing their full potential.

This book, *The Ultimate Guide to Making Money Online: Legitimate and Authentic Opportunities for Students, Women, Housewives, and Remote Workers*, is my endeavour to cut through the noise. It is a comprehensive roadmap designed to help you navigate the vast and dynamic world of online earning with confidence, clarity, and purpose.

**Why This Book Matters**

The idea of making money online often evokes skepticism—questions of legitimacy and authenticity are common, and rightly so. In this book, I focus exclusively on methods that are tried, tested, and genuinely impactful. From freelancing and blogging to e-commerce, affiliate marketing, and digital tutoring, every opportunity discussed here is rooted in integrity and practicality.

Moreover, this book is not just about earning money; it's about creating a balanced, sustainable lifestyle that aligns with your goals and values. Whether you want a part-time side hustle or aspire to transition into full-time online entrepreneurship, this guide provides the tools, tips, and insights to help you succeed.

**Who Is This Book For?**

This book is for:

- **Students** who wish to fund their education or gain financial independence.
- **Women and housewives** seeking to contribute to household income while managing personal responsibilities.
- **Remote workers and professionals** aiming to diversify their income streams or transition to flexible work.
- **Anyone with a vision** for a better future and the willingness to learn, adapt, and grow.

**What You'll Learn**

- How to identify the best online income opportunities suited to your skills and interests.
- Strategies to build scalable, sustainable income streams.
- Tips to avoid scams and navigate the digital landscape safely.
- Insights into the mindset and habits needed for long-term success.

**My Promise to You**

This book is not a get-rich-quick scheme or a collection of vague advice. It's a practical guide filled with actionable strategies, real-life examples, and proven methods. My aim is to empower you with the knowledge and confidence to take the first step—and every step after—toward financial freedom and independence.

# ACKNOWLEDGMENTS

This book is the result of years of learning, experimentation, and interaction with individuals across diverse backgrounds. To all my clients, colleagues, and readers who have inspired this work—thank you. Your stories of success and resilience continue to fuel my passion for guiding others on this journey.

As you turn these pages, I encourage you to approach the opportunities outlined here with an open mind and a determined spirit. Remember, the journey to online success begins not with perfection but with action. Take the leap, learn along the way, and trust that your efforts will lead to a brighter future.

Here's to your success!

**Zaheer Ahmed Shaik**
Certified General Health and Business Consultant
Author, Mentor, and Entrepreneur

# CONTENTS

**INTRODUCTION**

The internet has revolutionized the way we work, learn, and connect, creating countless opportunities for earning an income from the comfort of home. Whether you're a student balancing academic responsibilities, a homemaker looking to contribute financially, or simply someone exploring flexible ways to make money online, this book is your ultimate guide.

Gone are the days when earning a living required commuting to an office or adhering to rigid work schedules. Today, the digital economy empowers individuals to work remotely, set their own hours, and explore their passions while earning a sustainable income. However, navigating this vast and often confusing world of online work can be daunting, especially with the rise of scams and get-rich-quick schemes. This book is here to demystify the process, offering practical, legitimate, and authentic ways to make money online, tailored for diverse audiences.

Students often face financial constraints, making online earning a viable option to support education and personal needs. Women, particularly housewives, frequently seek opportunities that allow them to balance family responsibilities while contributing to household income. Others simply desire the freedom to work on their terms, free from traditional workplace constraints. Regardless of your background or goals, this book is designed to cater to your needs.

Each chapter is meticulously crafted to provide you with actionable insights, step-by-step guides, and inspiring success stories. You'll learn how to identify your skills, choose the right platform, and avoid common pitfalls. From freelancing and blogging to e-commerce and passive income strategies, every avenue is explored in detail, ensuring you have the knowledge and tools to succeed.

Moreover, this book emphasizes authenticity and sustainability. We focus on genuine methods that not only provide income but also allow you to build a reputation and create long-term value. You'll discover how to transform your passions and hobbies into income streams, ensuring that your online work aligns with your interests and strengths.

Whether you're looking to earn a side income, transition to full-time online work, or simply explore your options, *The Ultimate Guide to Making Money Online* is your trusted companion. Let this book be your roadmap to success, guiding you through the intricacies of the digital economy and helping you unlock your full potential.

# CHAPTER 1

## Introduction to the Digital Economy

### Understanding Online Opportunities

In the 21st century, the digital economy has become a cornerstone of global business, reshaping industries and creating new opportunities for individuals worldwide. From tech-savvy millennials to seniors embracing technology, the internet offers a wealth of possibilities to earn money legitimately and authentically.

The digital economy refers to economic activities powered by digital technologies. It encompasses e-commerce, digital marketing, online education, freelancing, and much more. Unlike traditional industries, the digital economy thrives on innovation, connectivity, and adaptability, making it accessible to people from diverse backgrounds and skill sets.

**Why Work Online?**

The appeal of online work lies in its flexibility, accessibility, and potential for growth. Unlike traditional jobs, online opportunities allow you to:

1. **Set Your Own Schedule:** Balance work with personal commitments.

2. **Work from Anywhere:** All you need is an internet connection.

3. **Start with Minimal Investment:** Many opportunities require little to no upfront cost.

4. **Explore Diverse Avenues:** From creative pursuits to technical roles, the possibilities are endless.

5. **Build Passive Income:** Create systems that generate income with minimal ongoing effort.

**Key Online Sectors to Explore**

1. **Freelancing Platforms:** Websites like Upwork, Fiverr, and Freelancer offer gigs in writing, design, programming, and more.

2. **Content Creation:** Platforms such as YouTube, TikTok, and Instagram enable creators to monetize their audience through ads, sponsorships, and merchandise.

3. **E-Commerce:** Amazon, Etsy, and Shopify empower entrepreneurs to sell physical and digital products.

4. **Online Education:** Platforms like Udemy and Teachable allow experts to share knowledge while earning.

5. **Affiliate Marketing:** Promote products and earn commissions through platforms like Amazon Associates or ShareASale.

**Challenges in the Digital Economy**

While opportunities are abundant, navigating the digital economy isn't without its challenges. Common obstacles include:

- **Competition:** Popular niches can be highly competitive.
- **Consistency:** Sustaining effort is key to long-term success.
- **Scams:** Staying vigilant against fraudulent schemes is critical.
- **Time Management:** Balancing online work with personal responsibilities requires discipline.

**The Importance of Authenticity**

As you embark on your online earning journey, authenticity is paramount. This means choosing legitimate platforms, building trust with your audience or clients, and delivering quality work. Authenticity not only enhances your reputation but also fosters sustainable income streams.

This chapter sets the foundation for your journey into the digital economy. As you progress through the book, you'll gain deeper insights into specific methods, learn how to maximize your earnings, and discover strategies to build a fulfilling and profitable online career.

# CHAPTER 2

## Preparing For Success

### Tools, Skills, and Mindset Needed to Thrive

Before diving into the vast array of online earning opportunities, it's essential to set a solid foundation for success. This involves equipping yourself with the right tools, developing key skills, and cultivating a mindset that ensures consistent growth. Online success is not solely about finding opportunities; it's about preparing yourself to seize them effectively.

## 1. Essential Tools for Online Work

While online earning often boasts minimal entry barriers, having the right tools is crucial for efficiency and professionalism.

### a. Hardware Requirements

- **Laptop or Desktop Computer:** Invest in a reliable device with good processing power and sufficient memory. A mid-range system is usually adequate for most tasks.

- **Smartphone:** Many platforms, especially social media and micro-tasking sites, work seamlessly on mobile devices.

- **Peripherals:** A good-quality headset, microphone, and webcam are essential for communication and content creation.

### b. Software and Applications

- **Communication Tools:** Apps like Zoom, Google Meet, and Microsoft Teams are critical for virtual meetings.

- **Project Management Tools:** Platforms like Trello, Asana, and Notion help organize tasks efficiently.

- **Cloud Storage:** Services like Google Drive, Dropbox, and OneDrive are indispensable for file sharing and backup.

- **Skill-Specific Software:** Depending on your chosen field, you may need tools like Adobe Creative Suite (for design), Grammarly (for writing), or QuickBooks (for accounting).

### c. Reliable Internet Connection

A stable, high-speed internet connection is non-negotiable. Aim for speeds that support video calls, file uploads, and seamless browsing.

## 2. Core Skills for Online Earning

Success in the online world is often skill-driven. While some skills are industry-specific, others are universally beneficial.

### a. Communication Skills

Clear and professional communication is key. Whether interacting with clients, negotiating contracts, or marketing your services, strong verbal and written skills set you apart.

### b. Time Management and Organization

Working online requires discipline. Use calendars, to-do lists, and time-blocking techniques to stay organized. Tools like Google Calendar or Clockify can be valuable.

### c. Technical Proficiency

- Basic computer literacy is a must.
- Learn to navigate platforms relevant to your niche. For instance, understanding WordPress is vital for blogging, while mastering Canva is helpful for social media management.

### d. Research and Analytical Skills

The ability to research trends, analyse data, and adapt to changes is a valuable asset. Online earning often involves staying ahead of market dynamics.

### e. Marketing and Branding

Promoting yourself or your services is crucial. Familiarize yourself with basic marketing concepts like SEO (Search Engine Optimization), email marketing, and social media advertising.

## 3. Developing the Right Mindset

The online earning journey can be unpredictable. A resilient and growth-oriented mindset helps navigate challenges effectively.

### a. Embrace Lifelong Learning

The digital landscape evolves rapidly. To stay competitive, commit to continuous learning through online courses, webinars, and tutorials. Platforms like Coursera, Udemy, and LinkedIn Learning are great resources.

### b. Cultivate Patience and Persistence

Online earning is rarely an overnight success. Whether you're building a blog or establishing a freelance career, persistence pays off.

### c. Stay Open to Feedback

Constructive criticism helps you improve. Seek feedback from clients, peers, or mentors and use it as a stepping stone for growth.

### d. Focus on Quality Over Quantity

Delivering high-quality work builds trust and reputation, which are invaluable in the digital economy. Avoid shortcuts and prioritize excellence.

### e. Manage Expectations

While success stories abound, they often come with years of effort. Be realistic about timelines and earnings as you start.

## 4. Building an Online Portfolio

Having a professional portfolio enhances credibility. Regardless of your chosen field, showcase your work online to attract clients or customers.

### a. Choose a Platform

- **Freelancers:** Build profiles on platforms like Upwork, Fiverr, or LinkedIn.
- **Creators:** Use personal websites, Behance, or Dribbble for design and artistic work.
- **E-commerce Sellers:** Develop a store on platforms like Shopify or Etsy.

### b. What to Include

- Your bio and skillset.
- Examples of your best work or projects.
- Testimonials or reviews from past clients.
- Contact information and links to social profiles.

### c. Keep it Updated

Regularly update your portfolio to reflect your latest skills and accomplishments.

## 5. Establishing a Productive Work Environment

### a. Create a Dedicated Workspace

Set up a quiet, comfortable space with minimal distractions. Personalize it to make it motivating and conducive to work.

### b. Set Boundaries

If you're working from home, communicate your work hours to family or roommates to avoid interruptions.

### c. Follow a Routine

Establish a daily routine to maintain consistency. Include time for work, learning, and relaxation to avoid burnout.

## 6. Financial Preparedness

### a. Start Small

Avoid investing large sums in tools or courses initially. Use free or trial versions until you're certain about their utility.

### b. Budget Wisely

Keep track of income and expenses. Use budgeting tools like Mint or YNAB to manage finances.

### `c. Plan for Taxes

Understand tax obligations for online earnings in your country. Set aside a portion of your income for taxes.

## 7. Building a Support Network

Surround yourself with like-minded individuals who share similar goals.

- Join online communities or forums in your niche.

- Network on LinkedIn or attend virtual events.

- Seek mentorship from experienced professionals.

## CONCLUSION

Preparing for success in the online earning world is a combination of equipping yourself with the right tools, honing essential skills, and cultivating a mindset that embraces growth. By laying this groundwork, you'll be well-positioned to navigate the challenges and opportunities that come your way. With preparation as your anchor, you're ready to explore the diverse paths to making money online, covered in the subsequent chapters.

# CHAPTER 3

## Freelancing

### Monetizing Your Skills and Expertise

Freelancing is one of the most accessible and versatile ways to make money online. It offers individuals the freedom to work on their terms while leveraging their existing skills. Whether you're a writer, designer, programmer, or marketer, freelancing provides opportunities to connect with clients worldwide, build a portfolio, and generate income. This chapter explores the ins and outs of freelancing, from identifying your niche to scaling your freelance business.

# 1. What is Freelancing?

Freelancing involves offering services to clients on a project basis, without being tied to a single employer. Freelancers typically work independently and are hired for specific tasks, projects, or ongoing contracts.

**Advantages of Freelancing:**

- **Flexibility:** Set your hours and work remotely.
- **Diverse Opportunities:** Work in various industries and explore different types of projects.
- **Control Over Income:** Choose projects that align with your desired pay scale and expertise.

**Challenges of Freelancing:**

- **Irregular Income:** Earnings can fluctuate based on project availability.
- **Client Management:** Balancing multiple clients and meeting deadlines can be demanding.
- **Competition:** Popular freelancing platforms are highly competitive, requiring strategic positioning.

# 2. Identifying Your Freelance Niche

The first step to freelancing success is identifying a niche that aligns with your skills and interests.

### a. Assess Your Skills and Expertise

Start by listing your core competencies. Examples include:

- Writing and Editing

- Graphic Design

- Programming and Software Development

- Digital Marketing

- Video Editing

## b. Research Market Demand

Check freelancing platforms like Upwork, Fiverr, and Toptal to understand which skills are in demand.

## c. Choose a Profitable Niche

Find the sweet spot between your skills, passions, and market demand. Specialized niches, like technical writing or UX design, often command higher rates.

## 3. Setting Up Your Freelance Business

Once you've identified your niche, it's time to establish your freelance presence.

## a. Build an Online Profile

Create accounts on popular freelancing platforms:

- **Upwork:** Great for diverse projects and long-term contracts.

- **Fiverr:** Ideal for offering specific services at set prices.

- **Toptal:** Focuses on high-quality projects for experienced freelancers.

**Profile Tips:**

- Use a professional photo.
- Write a compelling bio highlighting your expertise and accomplishments.
- List your skills and tools you're proficient in.
- Add a portfolio showcasing your best work.

## b. Create a Personal Website

A personal website acts as your digital portfolio and helps establish credibility. Include:

- A homepage introducing yourself.
- Portfolio samples categorized by project type.
- A contact form for inquiries.

## c. Define Your Services and Pricing

Be clear about the services you offer and the rates you charge. Consider:

- Hourly rates for ongoing tasks.
- Project-based pricing for fixed-scope work.

Research industry standards and adjust your rates as you gain experience.

## 4. Finding Freelance Clients

Attracting and retaining clients is a core aspect of freelancing.

## a. Freelancing Platforms

Bid on projects that match your skills. Write personalized proposals that explain how you can meet the client's needs.

## b. Social Media Marketing

Promote your services on platforms like LinkedIn, Twitter, and Instagram. Post engaging content that showcases your expertise.

## c. Networking

Join industry-specific groups, forums, and online communities. Engage in discussions and share insights to position yourself as an expert.

## d. Cold Outreach

Identify potential clients and reach out directly via email. Personalize your messages to explain how you can add value.

## 5. Delivering High-Quality Work

Securing a project is just the beginning; your success depends on delivering excellent results.

## a. Understand the Client's Needs

Communicate with clients to clarify project requirements, timelines, and deliverables. Use tools like Trello or Asana for project management.

### b. Meet Deadlines

Timely delivery builds trust and increases the likelihood of repeat business.

### c. Maintain Professionalism

- Respond to client queries promptly.
- Be open to feedback and revisions.
- Maintain a positive and collaborative attitude.

### d. Use High-Quality Tools

Invest in tools that enhance the quality of your work. For example, writers can use Grammarly for editing, while designers can rely on Adobe Creative Suite.

## 6. Scaling Your Freelance Career

As you gain experience, focus on growing your freelance business.

### a. Specialize in High-Demand Skills

Develop expertise in niche areas that offer premium pay rates. For instance:

- Blockchain development
- Conversion rate optimization
- Medical writing

## b. Build a Strong Portfolio

Expand your portfolio with high-quality, diverse projects. Use before-and-after examples to demonstrate the impact of your work.

## c. Increase Your Rates

As your experience and reputation grow, gradually raise your rates. Justify premium pricing with proven results and testimonials.

## d. Expand Your Network

Collaborate with other freelancers or agencies to secure larger projects. Attend virtual or in-person industry events to make connections.

## 7. Overcoming Freelancing Challenges

Freelancing is rewarding but not without its hurdles.

## a. Managing Irregular Income

Save a portion of each payment for lean periods. Diversify your client base to reduce reliance on a single source.

## b. Avoiding Burnout

Set boundaries to prevent overwork. Take breaks and prioritize self-care.

## c. Handling Difficult Clients

Be professional and assertive when dealing with unreasonable demands. If necessary, terminate contracts respectfully.

**d. Staying Updated**

Continuously learn new skills and adapt to industry trends. Subscribe to newsletters, blogs, and courses in your niche.

## CONCLUSION

Freelancing is a powerful way to monetize your skills while enjoying the flexibility of online work. With the right preparation, tools, and strategies, you can carve a successful freelance career that aligns with your goals and passions. Whether you're a beginner or an experienced professional, freelancing offers endless opportunities for growth and financial independence.

# CHAPTER 4

## Blogging

## Building a Profitable Content Platform

Blogging has evolved into a lucrative career for many, offering opportunities to share knowledge, express creativity, and earn a sustainable income. Whether you're a student, homemaker, or aspiring entrepreneur, creating a blog allows you to establish an online presence, connect with an audience, and generate revenue through various monetization strategies.

This chapter guides you through the process of starting a successful blog, from selecting a niche to implementing strategies for profitability.

## 1. Understanding Blogging as a Business

Blogging is more than just writing articles—it's about building a platform that provides value to readers and drives traffic, leading to monetization opportunities.

### Benefits of Blogging:

- **Creative Freedom:** Share your knowledge, experiences, and passions.
- **Scalability:** As your audience grows, your earning potential increases.
- **Diverse Income Streams:** Blogs can be monetized through ads, affiliate marketing, sponsored posts, and digital products.

### Challenges of Blogging:

- Consistent content creation requires dedication.
- Building a loyal audience takes time and effort.
- Monetization often depends on achieving substantial traffic.

## 2. Selecting a Niche for Your Blog

Choosing the right niche is critical to your blog's success.

### a. Evaluate Your Interests and Expertise

Write about topics you are passionate about or knowledgeable in, such as:

- Travel and Lifestyle
- Health and Fitness
- Technology and Gadgets

- Parenting and Education

## b. Analyse Market Demand

Use tools like Google Trends, Ubersuggest, or SEMrush to identify high-demand topics with manageable competition.

## c. Define Your Target Audience

Determine who your ideal readers are. Understand their needs, preferences, and pain points to create relevant content.

## d. Select a Profitable Niche

Some niches, such as personal finance, technology, and health, have higher earning potential due to lucrative advertising and affiliate opportunities.

## 3. Setting Up Your Blog

Starting a blog involves technical and creative steps to create a functional and appealing platform.

### a. Choose a Blogging Platform

Popular platforms include:

- **WordPress.org:** Offers flexibility and scalability.
- **Blogger:** Simple and free for beginners.
- **Wix or Squarespace:** User-friendly with drag-and-drop features.

## b. Select a Domain Name and Hosting

- **Domain Name:** Choose a name that reflects your niche and is easy to remember. Tools like Namecheap and GoDaddy are helpful for registration.

- **Hosting:** Reliable hosting providers include Bluehost, SiteGround, and HostGator.

## c. Design Your Blog

- Use a clean and responsive theme for easy navigation.

- Optimize for mobile users.

- Customize with your branding, including a logo and colour scheme.

## 4. Creating Engaging Content

Content is the backbone of any successful blog.

## a. Types of Blog Content

- **Informative Articles:** Provide solutions to common problems.

- **Listicles:** Organize information in a readable format (e.g., "Top 10 Travel Destinations").

- **How-To Guides:** Step-by-step instructions on specific topics.

- **Personal Stories:** Build relatability with your audience.

- **Product Reviews:** Offer detailed evaluations of products in your niche.

## b. Content Planning

- **Keyword Research:** Use tools like Ahrefs or Google Keyword Planner to identify trending keywords.

- **Content Calendar:** Schedule posts in advance to maintain consistency.

### c. Writing Tips

- Write in a conversational tone.

- Use headings, bullet points, and visuals to break up text.

- End posts with a call-to-action (CTA) to encourage reader engagement.

## 5. Driving Traffic to Your Blog

Without traffic, even the best content remains unnoticed.

### a. SEO Optimization

- **On-Page SEO:** Use target keywords in titles, headers, and meta descriptions.

- **Off-Page SEO:** Build backlinks by guest posting and networking with other bloggers.

- **Technical SEO:** Ensure fast loading speeds, secure (HTTPS) protocols, and mobile-friendliness.

### b. Leverage Social Media

Promote your blog on platforms like Pinterest, Instagram, and LinkedIn. Tailor your content to each platform's audience.

### c. Email Marketing

Build an email list using tools like Mailchimp or ConvertKit. Send newsletters with updates and exclusive content.

**d. Collaborate with Others**

Partner with fellow bloggers or influencers in your niche for cross-promotion.


## 6. Monetizing Your Blog

Once you have consistent traffic, implement monetization strategies to earn from your efforts.

### a. Advertising

- Sign up for ad networks like Google AdSense or Mediavine.
- Place banner ads strategically without disrupting user experience.

### b. Affiliate Marketing

- Promote products or services through affiliate programs like Amazon Associates or ShareASale.
- Write genuine reviews and include affiliate links.

### c. Sponsored Posts

- Collaborate with brands to write posts promoting their products.
- Ensure transparency by marking sponsored content clearly.

### d. Selling Digital Products

- Create e-books, courses, or templates related to your niche.
- Use platforms like Gumroad or Teachable for distribution.

## e. Memberships and Subscriptions

- Offer premium content or exclusive access to subscribers.
- Use tools like Patreon or WordPress membership plugins.

## 7. Maintaining and Growing Your Blog

A successful blog requires continuous effort to sustain and grow.

### a. Track Performance

- Use Google Analytics to monitor traffic and user behavior.
- Identify top-performing content and replicate its success.

### b. Update Old Content

Revise outdated posts with fresh information and new keywords.

### c. Engage with Readers

Respond to comments and emails promptly. Build a community around your blog.

### d. Stay Consistent

Post regularly to retain audience interest and boost SEO rankings.

# CONCLUSION

Blogging is a fulfilling and profitable online career when approached strategically. By selecting the right niche, creating valuable content, and implementing effective monetization methods, you can turn your blog into a sustainable income stream. Consistency, creativity, and audience engagement are the pillars of long-term success in blogging.

# CHAPTER 5

## Affiliate Marketing

## Turning Recommendations into Revenue

**Affiliate Marketing** has become one of the most effective ways to earn passive income online. This marketing model allows you to earn commissions by promoting products or services from other companies. By leveraging your platform, network, or content, you can turn simple recommendations into a steady stream of revenue. This chapter provides a comprehensive guide to affiliate marketing, from getting started to scaling your earnings.

# 1. What is Affiliate Marketing?

Affiliate marketing is a performance-based marketing model where you earn a commission for every sale, click, or lead generated through your referral links. It is a mutually beneficial arrangement:

- **Businesses** expand their reach without upfront advertising costs.
- **Affiliates** earn money by promoting products they trust.

## Key Players in Affiliate Marketing:

- **Merchants/Advertisers:** Businesses offering products or services.
- **Affiliates/Publishers:** Individuals promoting the merchant's products.
- **Customers:** The end users purchasing the product through affiliate links.
- **Affiliate Networks:** Platforms like ShareASale, CJ Affiliate, and ClickBank that connect merchants with affiliates.

# 2. Benefits of Affiliate Marketing

Affiliate marketing is popular for several reasons:

- **Low Startup Costs:** No need to create or stock products.
- **Flexibility:** Work anytime and anywhere.
- **Scalability:** Once content is live, it can generate revenue indefinitely.
- **Multiple Income Streams:** Promote products across various niches and networks.

# 3. Getting Started with Affiliate Marketing

To succeed in affiliate marketing, follow these essential steps:

## a. Choose a Niche

Selecting the right niche is crucial to your success. Choose one that:

- Aligns with your interests or expertise.
- Has a demand for affiliate products (e.g., technology, fitness, personal finance).
- Has a good balance of competition and opportunity.

## b. Join Affiliate Programs

Look for programs that offer:

- Competitive commission rates.
- High-quality products with proven demand.
- Reliable tracking and payment systems.

Popular affiliate programs include:

- **Amazon Associates:** Ideal for beginners, offering a vast product catalogue.
- **ShareASale:** A diverse network with various merchants.
- **CJ Affiliate:** High-commission programs in multiple industries.

## c. Build a Platform

Affiliate marketing works best when you have an established platform:

- **Blog or Website:** Ideal for detailed reviews and evergreen content.

- **Social Media Channels:** Effective for quick promotions and engaging content.
- **YouTube:** Perfect for tutorials, unboxings, and product demonstrations.

### d. Create Quality Content

Successful affiliate marketers focus on creating value-driven content:

- Product reviews and comparisons.
- Tutorials and how-to guides featuring affiliate products.
- Personal testimonials and case studies.

## 4. Crafting Effective Affiliate Content

Content is king in affiliate marketing. Your content should be authentic, engaging, and tailored to your audience.

### a. Understand Your Audience

Identify your audience's needs, pain points, and preferences. For example:

- A tech-savvy audience may prefer in-depth product comparisons.
- A fitness-focused audience might appreciate workout gear recommendations.

### b. Use Storytelling

Weave personal experiences into your content. For example:

- "I struggled with productivity until I discovered this planner…"

### c. Incorporate Visuals

Visual elements such as images, infographics, and videos increase engagement. Tools like Canva and Adobe Spark can help you create appealing visuals.

### d. Add Clear CTAs

Encourage readers to take action with clear calls-to-action (CTAs):

- "Click here to get started with [Product Name] today!"
- "Check out [Brand Name] for the best deals on [Product]."

## 5. Promoting Your Affiliate Links

Strategic promotion ensures your affiliate links reach the right audience.

### a. SEO Optimization

- Use relevant keywords to improve your content's search engine visibility.
- Optimize titles, headers, and meta descriptions with primary keywords.

### b. Email Marketing

Build an email list and share affiliate products through newsletters. Personalize your emails for better engagement.

### c. Social Media Campaigns

Leverage platforms like Instagram, Pinterest, and Twitter to share affiliate links. Use engaging visuals and hashtags to expand your reach.

**d. Paid Advertising**

Use platforms like Google Ads or Facebook Ads to target specific audiences. Be cautious of ROI and start with a small budget.

## 6. Tracking and Analysing Performance

Monitoring your affiliate marketing campaigns helps optimize performance.

**a. Metrics to Track:**

- Click-through rate (CTR).
- Conversion rate.
- Average order value (AOV).
- Total revenue generated.

**b. Tools for Analysis:**

- Google Analytics: Understand audience behaviour on your website.
- Affiliate dashboards: Review performance metrics provided by networks.

**c. Experiment and Optimize:**

- A/B test CTAs, content formats, and promotional strategies.
- Focus on products or niches with higher conversion rates.

## 7. Overcoming Common Challenges

Affiliate marketing is rewarding but not without challenges.

### a. Fierce Competition

Stand out by offering unique insights or targeting a specific audience segment.


### b. Low Conversion Rates

Build trust by promoting only products you believe in. Provide detailed, unbiased reviews.


### c. Algorithm Changes

Stay updated with search engine and social media platform policies. Diversify traffic sources to minimize impact.


## 8. Scaling Your Affiliate Marketing Business

To grow your affiliate marketing business, consider these strategies:


### a. Expand Your Content Portfolio

Cover more topics within your niche or explore related niches.


### b. Diversify Affiliate Programs

Don't rely solely on one program. Explore high-paying affiliate opportunities.


### c. Invest in Paid Tools

Premium tools like SEMrush or Ahrefs can help you improve keyword research and SEO.

**d. Automate Processes**

Use automation tools like Hootsuite for social media scheduling and ActiveCampaign for email marketing.

## CONCLUSION

Affiliate marketing is a powerful way to generate income by sharing products and services you believe in. By choosing the right niche, creating high-quality content, and strategically promoting your links, you can build a sustainable and scalable business. Whether you're a beginner or an experienced marketer, affiliate marketing offers endless opportunities to turn recommendations into revenue.

# CHAPTER 6

## Online Tutoring

### Sharing Knowledge and Earning Money

Online tutoring has revolutionized education, enabling individuals to share their knowledge and skills with students worldwide while earning a stable income. As a tutor, you can teach subjects, offer professional training, or provide personalized guidance in various disciplines. This chapter explores the tools, strategies, and opportunities available to build a successful online tutoring career.

**1. The Growing Demand for Online Tutoring**

The rise of e-learning platforms and the need for flexible education options have made online tutoring a lucrative field. Factors driving this demand include:

- **Global Accessibility:** Students from different time zones can access lessons.
- **Flexible Schedules:** Suitable for learners and tutors with varying time commitments.
- **Personalized Learning:** One-on-one sessions cater to individual needs.

**Key Market Segments:**
- K-12 Education
- Test Preparation (SAT, GRE, GMAT)
- Language Learning (English, Spanish, etc.)
- Professional Skills (coding, graphic design, etc.)

**2. Identifying Your Expertise**

Successful online tutors leverage their strengths and passions. Follow these steps to define your niche:

**a. Assess Your Skills and Qualifications**
- What are your academic or professional qualifications?
- Do you have unique skills like music, art, or coding?

## b. Understand Market Demand

Research popular tutoring categories. Platforms like Preply, Tutor.com, and Wyzant list high-demand subjects.

## c. Specialize in a Niche

Focus on a specific area to stand out. For example:

- Math tutoring for high school students.
- Conversational English for non-native speakers.

## 3. Setting Up Your Online Tutoring Business

Before teaching your first lesson, prepare the essentials:

## a. Choose a Tutoring Model

- **One-on-One Tutoring:** Personalized sessions with higher earning potential.
- **Group Classes:** Teach multiple students simultaneously for scalability.
- **Pre-Recorded Courses:** Earn passive income by creating downloadable lessons.

## b. Select a Platform

Decide where to host your lessons:

- **Freelance Platforms:** VIPKid, Preply, Chegg Tutors.
- **Dedicated Websites:** Create a personal tutoring site using tools like WordPress or Teachable.
- **Marketplaces:** Udemy and Skillshare for pre-recorded courses.

## c. Invest in Equipment

- High-quality webcam and microphone for clear communication.
- Stable internet connection to avoid disruptions.
- Interactive tools like a digital writing tablet for math or art lessons.

## 4. Crafting an Effective Lesson Plan

Planning is crucial for delivering value-packed lessons.

### a. Define Learning Objectives

Break down what students should achieve by the end of each session. For example:

- Understand algebraic equations.
- Master basic conversational phrases in French.

### b. Structure Your Sessions

- **Introduction:** Brief recap of previous lessons or an icebreaker.
- **Core Content:** Use visuals, examples, and interactive methods to teach.
- **Practice:** Assign exercises to reinforce concepts.
- **Conclusion:** Summarize key points and outline next steps.

### c. Use Engaging Materials

- Presentations (Google Slides, PowerPoint).
- Videos (YouTube tutorials, recorded demonstrations).
- Worksheets and quizzes for practice.

## 5. Building a Strong Online Presence

To attract students, establish yourself as a credible and approachable tutor.

### a. Create a Professional Profile

Include:

- Your qualifications and experience.
- A clear photo and an introduction video.
- Testimonials or reviews from past students.

### b. Showcase Your Expertise

- Share educational blog posts or videos on social media.
- Offer free webinars or sample lessons.

### c. Network with Other Educators

Join online forums, LinkedIn groups, or Facebook communities to connect with like-minded professionals and potential students.

## 6. Setting Your Pricing

Determine competitive rates based on your niche, experience, and the market.

### Factors to Consider:

- Demand for your subject.
- Duration of sessions (hourly, weekly packages).

- Additional materials provided (study guides, recorded lectures).

**Pricing Strategies:**

- Start with promotional rates to build a client base.
- Increase prices as you gain experience and positive reviews.

# 7. Delivering Outstanding Lessons

Student satisfaction is key to success in tutoring.

## a. Foster Engagement

- Use interactive tools like Zoom's whiteboard feature or Kahoot quizzes.
- Encourage questions and active participation.

## b. Be Patient and Supportive

Every student learns at their own pace. Celebrate small achievements to motivate them.

## c. Provide Feedback

- Highlight strengths and areas for improvement.
- Offer actionable advice to help students progress.

# 8. Expanding and Scaling Your Tutoring Business

Once you've established a solid foundation, consider scaling up.

## a. Diversify Your Offerings

- Add more subjects or advanced levels.

- Create supplementary resources like e-books or practice tests.

## b. Automate Administrative Tasks

Use scheduling tools like Calendly and payment systems like PayPal to streamline operations.

## c. Collaborate with Institutions

Partner with schools, coaching centres, or online learning platforms to expand your reach.

## CONCLUSION

Online tutoring is a fulfilling and profitable way to share your expertise while helping others achieve their learning goals. Whether you're teaching a language, preparing students for exams, or sharing professional skills, the potential for growth and income is substantial. By focusing on quality, consistency, and building strong relationships with your students, you can create a rewarding career as an online tutor.

# CHAPTER 7

## E-commerce and Dropshipping

### Selling Products Online

The world of e-commerce has opened up endless opportunities for entrepreneurs to sell products globally without the need for physical stores. Among the various e-commerce models, dropshipping has gained immense popularity for its low-risk approach. In this chapter, we'll explore the basics of e-commerce, the essentials of setting up an online store, and the step-by-step process to start a successful dropshipping business.

## 1. The Rise of E-commerce and Dropshipping

E-commerce has transformed the way people shop, enabling businesses to reach customers worldwide. Dropshipping, in particular, has made it easier for individuals to start an online business without worrying about inventory or logistics.

### Key Advantages of E-commerce:

- Global reach and 24/7 availability.
- Low overhead costs compared to physical stores.
- Easy scalability to expand product offerings.

### What Makes Dropshipping Unique?

- **No Inventory Management:** Products are shipped directly from suppliers to customers.
- **Low Startup Costs:** No need to invest in stock upfront.
- **Flexibility:** Operate the business from anywhere with an internet connection.

## 2. Choosing the Right E-commerce Business Model

E-commerce offers various business models, including:

### a. Traditional Online Store:

You buy inventory in bulk, store it, and ship it to customers. Ideal for businesses with unique or branded products.

## b. Dropshipping:

You act as a middleman, listing products on your website while a supplier fulfils the orders. Great for beginners due to its simplicity.

## c. Print-on-Demand:

Sell custom-designed products like T-shirts or mugs, with orders printed and shipped only when purchased.

## d. Digital Products:

Sell non-physical items like e-books, courses, or software. High-profit margins as there are no shipping or manufacturing costs.

## 3. Finding the Right Products to Sell

Choosing the right products is critical for success in e-commerce and dropshipping.

### a. Look for High-Demand Niches:

- Health and fitness (e.g., yoga mats, resistance bands).

- Pet supplies (e.g., chew toys, grooming kits).

- Tech gadgets and accessories.

### b. Avoid Oversaturated Markets:

Research competitors and choose less crowded niches with sufficient demand.

### c. Evaluate Profit Margins:

Focus on products that offer a good balance between cost and selling price.

### Tools for Product Research:

- Google Trends: Analyse product popularity over time.

- AliExpress: Find trending products for dropshipping.

- Jungle Scout: Research profitable products on Amazon.

## 4. Setting Up Your E-commerce Store

Launching an e-commerce store involves several key steps:

### a. Choose a Platform

Select a platform based on your needs and technical expertise. Popular options include:

- **Shopify:** User-friendly and ideal for dropshipping.
- **WooCommerce:** A customizable plugin for WordPress.
- **BigCommerce:** Scalable and feature-rich.

### b. Register a Domain Name

Choose a domain name that reflects your brand and niche. Keep it simple, memorable, and easy to spell.

### c. Design Your Store

- Use visually appealing themes or templates.
- Highlight featured products on the homepage.
- Optimize for mobile devices.

### d. Add Payment Gateways

Integrate secure payment options like PayPal, Stripe, or credit card processors to make transactions seamless.

## 5. Understanding Dropshipping Logistics

Dropshipping eliminates the hassle of inventory management but requires careful coordination with suppliers.

### a. Partnering with Reliable Suppliers

- Platforms like Oberlo, Spocket, and AliExpress connect you with dropshipping suppliers.
- Test supplier reliability by placing sample orders.

### b. Product Listings and Pricing

- Write detailed and compelling product descriptions.

- Use high-quality images and videos to showcase products.

- Price competitively while maintaining a healthy profit margin.

### c. Order Fulfilment Process

- When a customer places an order, you forward the details to your supplier.

- The supplier ships the product directly to the customer.

## 6. Marketing Your Online Store

Driving traffic to your e-commerce store is crucial for generating sales.

### a. Social Media Marketing

- Use Instagram and Pinterest for visually appealing product promotions.

- Run targeted ads on Facebook to reach specific demographics.

### b. Search Engine Optimization (SEO)

- Optimize product titles and descriptions with relevant keywords.

- Write blog posts that attract organic traffic to your site.

### c. Email Marketing

- Build an email list using pop-up sign-up forms.

- Send personalized offers and product recommendations to subscribers.

### d. Influencer Collaborations

Partner with influencers in your niche to promote your products.

## 7. Managing Challenges in E-commerce

While e-commerce and dropshipping are rewarding, challenges are inevitable:

### a. Long Shipping Times:

- Use local suppliers or warehouses to reduce delivery times.
- Set clear expectations with customers about shipping durations.

### b. Customer Service Issues:

- Respond promptly to inquiries and complaints.
- Offer refunds or replacements for defective products.

### c. Maintaining Competitive Pricing:

- Regularly review competitor prices.
- Negotiate better deals with suppliers for bulk orders.

## 8. Scaling Your E-commerce Business

Once your store gains traction, focus on growth strategies:

### a. Expand Product Lines:

Introduce complementary products to upsell or cross-sell.

**b. Automate Processes:**

- Use tools like Zapier or Shopify Flow for order processing.

- Automate marketing campaigns with email tools like Klaviyo.

**c. Invest in Paid Advertising:**

Scale your ad campaigns on platforms like Google Ads and Facebook.

## CONCLUSION

E-commerce and dropshipping offer endless opportunities for individuals to earn money online. By selecting the right products, building a user-friendly store, and leveraging effective marketing strategies, you can create a thriving business. The flexibility and scalability of this model make it an excellent choice for students, homemakers, and anyone looking to work from home.

# CHAPTER 8

## Virtual Assistance

### Supporting Businesses from Anywhere

In today's interconnected digital landscape, businesses and entrepreneurs often seek reliable support to manage their operations. Virtual assistance has emerged as a thriving opportunity, enabling skilled individuals to offer administrative, creative, and technical services from the comfort of their homes. This chapter explores the essential skills, tools, and strategies needed to succeed as a virtual assistant (VA), along with insights into finding clients and scaling your services.

# 1. The Role of a Virtual Assistant

A virtual assistant provides remote support to clients, handling a variety of tasks that help businesses operate smoothly. The scope of work often includes:

- Administrative tasks (scheduling, email management, data entry).
- Creative services (content writing, graphic design).
- Technical support (website management, SEO optimization).
- Specialized roles (social media management, bookkeeping).

## Benefits of Being a Virtual Assistant:

- Flexible work hours and location.
- Opportunities to work with diverse industries.
- Low startup costs and minimal equipment requirements.

# 2. Skills Required to Excel as a Virtual Assistant

To thrive in the virtual assistance field, focus on developing these core skills:

### a. Time Management and Organization

- Ability to prioritize tasks effectively.
- Familiarity with project management tools like Trello or Asana.

### b. Communication Skills

- Clear and professional communication via email, chat, and video calls.
- Proficiency in tools like Zoom, Microsoft Teams, or Slack.

## c. Technical Proficiency

- Knowledge of office applications like Microsoft Office and Google Workspace.
- Familiarity with specific tools relevant to clients, such as Canva for graphic design or Hootsuite for social media management.

## d. Adaptability and Problem-Solving

- Quick learning and adjustment to new tools or industries.
- Creative approaches to overcoming challenges.

# 3. Setting Up Your Virtual Assistant Business

## a. Define Your Niche

Specializing in a particular area can set you apart from competitors. Consider niches like:

- Real estate virtual assistance.
- E-commerce support (order processing, inventory management).
- Executive assistance for CEOs and managers.

## b. Create a Professional Online Presence

- Build a website or LinkedIn profile showcasing your skills and services.
- Include testimonials or case studies if available.

## c. Invest in Essential Tools

- A reliable computer and high-speed internet.

- Headphones with a microphone for virtual meetings.
- Task management and time-tracking software like Toggl.

### d. Legal and Financial Preparation

- Register your business if necessary and comply with local regulations.
- Use invoicing tools like QuickBooks or Wave for professional billing.

## 4. Finding Clients

Securing clients is crucial to building a successful VA career.

### a. Freelance Platforms

- Use websites like Upwork, Fiverr, and Freelancer to connect with clients.
- Optimize your profile with keywords that highlight your expertise.

### b. Networking

- Join professional groups on LinkedIn or Facebook.
- Attend virtual industry events to connect with potential clients.

### c. Cold Outreach

- Research businesses or entrepreneurs who may need your services.
- Send personalized emails offering your expertise and explaining how you can add value.

**d. Referrals and Testimonials**

- Encourage satisfied clients to recommend your services.
- Request testimonials to build credibility.

## 5. Delivering Outstanding Service

**a. Understand Client Expectations**

- Conduct an onboarding session to clarify tasks, tools, and timelines.
- Create a contract detailing the scope of work, payment terms, and deadlines.

**b. Maintain Clear Communication**

- Provide regular updates on progress.
- Address concerns promptly and professionally.

**c. Go the Extra Mile**

- Suggest process improvements or new tools that can benefit the client.
- Offer to take on additional responsibilities if it aligns with your skills.

## 6. Challenges and How to Overcome Them

**a. Managing Multiple Clients**

- Use project management tools to stay organized.
- Set clear boundaries to avoid overcommitment.

## b. Inconsistent Workload

- Build a pipeline of potential clients to ensure consistent income.
- Offer retainer agreements for long-term projects.

## c. Technical Issues

- Invest in backup systems like cloud storage and alternate internet connections.
- Stay updated with software and tools used in your field.

## 7. Scaling Your Virtual Assistance Business

### a. Diversify Your Services

- Add high-demand skills like email marketing or graphic design.
- Offer packages combining multiple services.

### b. Build a Team

- Hire subcontractors to handle specific tasks as your client base grows.
- Use collaboration tools like ClickUp or Monday.com to manage teamwork.

### c. Automate Routine Tasks

- Use tools like Zapier to streamline repetitive processes.
- Automate email responses or appointment scheduling.

### d. Upsell and Cross-Sell

- Offer additional services like creating templates or managing client databases.

- Identify areas where your existing clients may need further support.

## CONCLUSION

Virtual assistance is a versatile and rewarding way to earn money online while supporting businesses worldwide. By honing your skills, building a strong online presence, and consistently delivering value to your clients, you can establish a thriving VA career. Whether you start small or aim to grow into a full-fledged agency, the opportunities in this field are limitless.

# CHAPTER 9

## Creative Ventures

### Graphic Design, Photography, and More

The digital revolution has unlocked countless opportunities for creatives to turn their passions into profitable careers. Graphic design, photography, and other artistic pursuits are not only fulfilling but can also serve as sustainable sources of income when approached strategically. This chapter delves into the pathways for monetizing creative skills, building a brand, and scaling your ventures in a competitive marketplace.

## 1. The Scope of Creative Careers in the Digital Age

Creative professions are in demand across various industries, including marketing, entertainment, publishing, and e-commerce. Businesses rely on visual storytelling and high-quality content to stand out, creating a robust market for skilled individuals in graphic design, photography, videography, and beyond.

### Key Advantages of Creative Careers Online:

- Flexibility to work on a freelance basis or as a full-time professional.
- Opportunities to collaborate with global clients.
- Diverse income streams, including direct client work, licensing, and digital product sales.

## 2. Getting Started: Identifying Your Creative Niche

To establish yourself in the creative industry, begin by identifying your niche. Some popular options include:

### a. Graphic Design:

- Logo creation and brand identity design.
- Social media graphics and web banners.
- UI/UX design for apps and websites.

### b. Photography:

- Stock photography for platforms like Shutterstock and Getty Images.
- Event photography (weddings, parties, corporate events).
- Product and lifestyle photography for e-commerce brands.

## c. Videography:

- Video editing for content creators and businesses.
- Short film and promotional video production.

## d. Specialized Art:

- Illustration and digital painting for books, games, and advertising.
- Custom art commissions and prints.

## 3. Building Your Skillset

### a. Learn the Basics

- Take online courses on platforms like Coursera, Udemy, or Skillshare.
- Study foundational concepts such as color theory, composition, and typography.

### b. Master the Tools

- **Graphic Design:** Adobe Photoshop, Illustrator, Canva, or Figma.
- **Photography:** Adobe Lightroom, Photoshop, or Capture One for editing.
- **Videography:** Adobe Premiere Pro, Final Cut Pro, or DaVinci Resolve.

### c. Practice and Portfolio Development

- Undertake personal projects to refine your skills.
- Create a portfolio showcasing your best work to attract clients.

## 4. Monetizing Graphic Design Skills

### a. Freelancing Platforms

- Join marketplaces like Fiverr, Upwork, or 99designs to find clients.
- Offer specialized services like social media templates or packaging design.

### b. Digital Products

- Sell templates, presets, or digital art on platforms like Creative Market or Gumroad.
- Create an online store to market custom design assets.

### c. Collaborative Projects

- Partner with small businesses to design their branding materials.
- Collaborate with other creatives to expand your network and visibility.

## 5. Earning Through Photography

### a. Stock Photography

- Upload high-quality images to stock platforms and earn royalties per download.
- Focus on themes like nature, travel, or lifestyle that have high demand.

### b. Direct Client Work

- Market your services to local businesses or individuals.
- Offer photography packages for events, products, or portraits.

## c. Selling Prints

- Open an online store or partner with art galleries to sell prints of your work.

- Experiment with limited-edition series to increase exclusivity and value.

## 6. Exploring Other Creative Ventures

### a. Illustration and Digital Art

- Monetize through commissions for custom pieces.

- Offer prints and merchandise featuring your art on sites like Redbubble or Society6.

### b. Animation and Motion Graphics

- Create explainer videos or animated ads for businesses.

- Market your work to content creators and production companies.

### c. Writing and Creative Content

- Write scripts, stories, or copy for businesses and entertainment projects.

- Self-publish books or e-books if you have a flair for storytelling.

## 7. Marketing Your Creative Services

### a. Build a Professional Online Presence

- Create a website or portfolio showcasing your work and expertise.

- Use platforms like Behance, Dribbble, or Instagram to connect with potential clients.

### b. Social Media Promotion

- Share behind-the-scenes content and finished projects to engage your audience.
- Collaborate with influencers or brands to gain exposure.

### c. Networking and Referrals

- Attend virtual and physical industry events.
- Build relationships with satisfied clients who can recommend your services.

## 8. Overcoming Challenges in the Creative Industry

### a. Managing Client Expectations

- Set clear deliverables and timelines in project agreements.
- Communicate effectively to avoid misunderstandings.

### b. Pricing Your Services

- Research market rates and price your work competitively.
- Offer tiered pricing packages to cater to different budgets.

### c. Staying Motivated

- Join creative communities for support and inspiration.
- Take breaks and engage in personal projects to prevent burnout.

## 9. Scaling Your Creative Business

### a. Diversify Your Income Streams

- Combine client work with passive income from digital products or print sales.
- Expand into related fields, such as videography if you're a photographer.

### b. Leverage Technology

- Use AI tools like Adobe Sensei for faster and smarter design workflows.
- Automate routine tasks, such as client invoicing, to save time.

### c. Build a Team

- Hire assistants or collaborators to handle tasks like editing or administration.
- Focus on high-level creative direction as your business grows.

## CONCLUSION

Creative ventures offer endless possibilities for individuals to monetize their artistic skills and passion. Whether you're designing stunning visuals, capturing unforgettable moments, or crafting compelling stories, the key lies in honing your craft, building a strong brand, and delivering consistent value to your clients. By embracing opportunities and adapting to industry trends, you can create a fulfilling and profitable career in the creative field.

# CHAPTER 10

# Content Creation

## YouTube, Podcasting, and Social Media Platforms

Content creation has revolutionized how individuals share ideas, entertain, educate, and inspire audiences while monetizing their creativity. Platforms like YouTube, podcasts, and social media channels have democratized access to global audiences, making it possible for anyone to turn their passion into a lucrative business. This chapter dives into the strategies, tools, and techniques for building a successful content creation career.

## 1. The Power of Content Creation

Content creation is a multi-billion-dollar industry, driven by the demand for engaging, relatable, and informative material. It spans across multiple formats:

- **Video content** (YouTube, TikTok).

- **Audio content** (podcasts).

- **Written and visual content** (Instagram, blogs).

Businesses, brands, and audiences rely on creators to provide value, entertainment, or solutions to their problems. By tapping into this market, creators can build both influence and income.

## 2. Choosing the Right Platform

Your choice of platform depends on your skills, target audience, and content format:

### a. YouTube

- Ideal for video content like tutorials, vlogs, and reviews.

- Strong monetization potential through ads, sponsorships, and memberships.

### b. Podcasting

- Perfect for conversational or educational content.

- Monetization options include sponsorships, donations, and premium episodes.

### c. Social Media Platforms

- Instagram, TikTok, and Facebook cater to visual and short-form content.

- Opportunities for influencer partnerships and selling digital products.

**d. Multi-Channel Strategy**

- Repurpose content across multiple platforms to maximize reach.

- For example, create a YouTube video, share clips on Instagram, and discuss the topic on your podcast.

## 3. Finding Your Niche

Your niche determines your audience and sets the tone for your content. To find the right niche:

**a. Identify Your Passion and Expertise**

- What topics do you enjoy exploring?

- What knowledge or skills can you share with others?

**b. Research Market Demand**

- Use tools like Google Trends or YouTube Analytics to identify trending topics.

- Analyse competitors to find gaps you can fill.

**c. Experiment and Refine**

- Test different formats and topics to see what resonates with your audience.

- Listen to feedback and adjust your content strategy accordingly.

## 4. Creating High-Quality Content

To stand out in a saturated market, focus on creating content that is both engaging and professional.

### a. Plan Your Content

- Develop a content calendar to maintain consistency.
- Script or outline your videos and podcasts to stay organized.

### b. Invest in Equipment

- For YouTube: Use a high-resolution camera, a good microphone, and lighting equipment.
- For Podcasting: A quality microphone, headphones, and editing software like Audacity or GarageBand are essential.

### c. Optimize Production Quality

- Edit your videos and audio to remove errors and enhance clarity.
- Use software like Adobe Premiere Pro, Final Cut Pro, or Canva for professional editing.

### d. Engage Your Audience

- Use storytelling techniques to connect emotionally.
- Include calls to action (e.g., like, comment, share, subscribe).

## 5. Growing Your Audience

Audience growth requires a mix of consistency, promotion, and community building.

### a. Consistency Is Key

- Post regularly to keep your audience engaged.
- Stick to a predictable schedule to build anticipation.

### b. Leverage Social Media

- Share teasers and behind-the-scenes content to drive traffic to your main platform.
- Use trending hashtags and participate in relevant conversations.

### c. Collaborate with Others

- Partner with other creators to expand your reach.
- Guest appearances on podcasts or cross-promotions can attract new followers.

### d. Interact with Your Community

- Respond to comments and messages to build trust and loyalty.
- Host live sessions or Q&A events to connect with your audience.

## 6. Monetizing Your Content

Monetization strategies vary depending on the platform:

### a. YouTube

- **Ad Revenue:** Earn through Google AdSense based on views and clicks.
- **Channel Memberships and Super Chats:** Offer exclusive content to paying subscribers.
- **Sponsorships:** Partner with brands for paid promotions.

### b. Podcasting

- **Sponsorships:** Promote products or services during episodes.
- **Listener Support:** Use platforms like Patreon for crowdfunding.
- **Merchandise Sales:** Sell branded merchandise or related products.

### c. Social Media

- **Affiliate Marketing:** Earn commissions by promoting products.
- **Brand Partnerships:** Collaborate with companies to create sponsored posts.
- **Digital Products:** Sell e-books, courses, or presets directly to followers.

## 7. Overcoming Challenges

Content creation is rewarding but comes with its challenges:

### a. Managing Time

- Batch-create content to streamline your workflow.
- Use scheduling tools like Buffer or Hootsuite to automate posting.

### b. Staying Relevant

- Stay updated on trends and platform algorithms.
- Experiment with new content formats to keep your audience engaged.

### c. Dealing with Criticism

- Focus on constructive feedback to improve.
- Develop resilience to handle negativity without losing motivation.

## 8. Scaling Your Content Creation Business

### a. Diversify Income Streams

- Combine multiple monetization strategies to stabilize income.
- Explore creating your products, like courses or books, based on your expertise.

### b. Build a Team

- Hire editors, graphic designers, or social media managers to handle specific tasks.
- Delegate responsibilities to focus on creative aspects.

### c. Expand Your Brand

- Launch related ventures like an e-commerce store or public speaking engagements.
- Build a personal brand to open new opportunities beyond content creation.

## CONCLUSION

Content creation is not just a hobby but a powerful career path that allows individuals to express their creativity, share knowledge, and generate income. By mastering the art of storytelling, leveraging technology, and engaging with audiences, you can turn platforms like YouTube, podcasts, and social media into thriving businesses. Start small, stay consistent, and let your passion guide you as you build a sustainable content creation empire.

# CHAPTER 11

## Surveys and Microtasks

### Earning Through Small Efforts

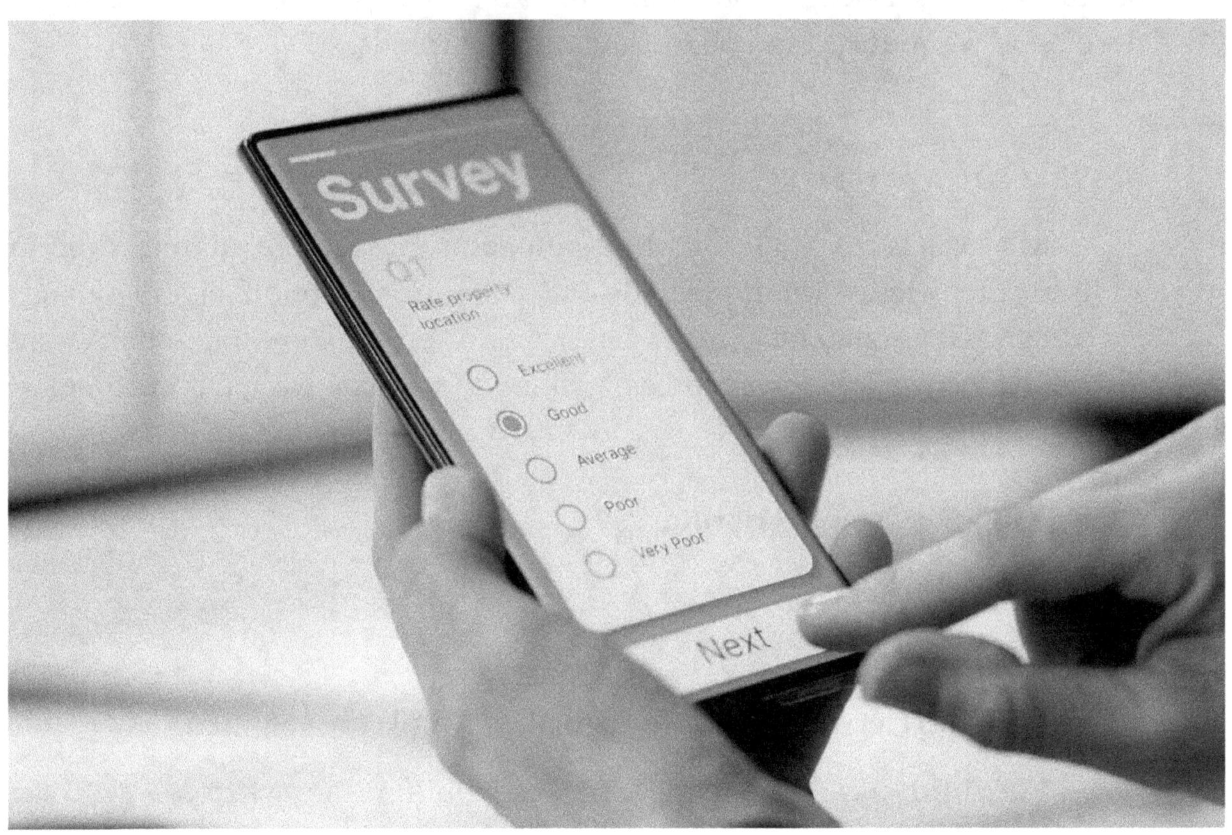



Not all online income methods require extensive skills, expertise, or time. Platforms offering surveys and microtasks provide an accessible entry point for anyone willing to dedicate effort to earning money online. These methods are particularly appealing for students, homemakers, and individuals looking for supplemental income, as they require minimal commitment and can be done at your own pace.

This chapter explores the world of surveys and microtasks, the platforms to use, tips for maximizing earnings, and how to avoid potential pitfalls.

## 1. Understanding Surveys and Microtasks

### a. What Are Online Surveys?

Online surveys are questionnaires designed by companies, marketers, or researchers to gather consumer opinions, preferences, or feedback on products and services. Companies rely on this data to refine strategies, develop new products, and understand market trends.

### b. What Are Microtasks?

Microtasks are small, simple assignments that require minimal time and skills to complete. Examples include data entry, transcription, tagging images, or categorizing content. These tasks are often broken into chunks and distributed across large workforces, making them ideal for quick income.

## 2. Why Choose Surveys and Microtasks?

### Advantages:

- **Low Barrier to Entry:** No specialized skills are required.
- **Flexibility:** Work on tasks or surveys anytime, anywhere.
- **Supplemental Income:** Ideal for earning additional money without long-term commitment.

### Limitations:

- **Limited Earnings:** Tasks typically pay small amounts, so income depends on volume.
- **Repetitive Nature:** Many tasks are monotonous.
- **Time-Intensive for Low Pay:** Some tasks may not justify the effort.

## 3. Platforms for Online Surveys

Numerous platforms connect individuals with companies offering paid surveys. Popular and trusted ones include:

### a. Swagbucks

- Offers points for surveys, watching videos, and shopping online.
- Points can be redeemed for cash or gift cards.

### b. Toluna

- Allows users to share opinions in exchange for rewards.
- Includes polls and product testing opportunities.

### c. Survey Junkie

- Focused exclusively on surveys.
- Users earn points redeemable via PayPal or gift cards.

### d. Pinecone Research

- A high-paying and reputable survey platform.
- Surveys typically involve evaluating new products.

## 4. Platforms for Microtasks

Microtask platforms offer quick assignments and payouts. Some of the most reliable platforms include:

### a. Amazon Mechanical Turk (MTurk)

- Tasks include data validation, transcription, and image tagging.
- Suitable for earning in small increments consistently.

### b. Clickworker

- Offers tasks like writing, research, and categorization.
- Payments are made via PayPal or SEPA transfer.

### c. Appen

- Focused on linguistic and AI-related tasks, such as speech annotation.
- Higher pay for specialized tasks compared to general platforms.

### d. Microworkers

- Short tasks like following social media accounts, verifying data, or watching videos.
- Easy to use for beginners.

## 5. How to Maximize Earnings from Surveys and Microtasks

### a. Choose Reputable Platforms

- Research platforms to ensure they have a history of paying users.
- Avoid scams or sites with high payout thresholds.

### b. Optimize Your Profile

- Complete profiles on survey sites to receive relevant and higher-paying surveys.
- Update demographic information regularly.

### c. Work Smart

- Focus on high-paying tasks and surveys.
- Avoid tasks that are time-consuming relative to their payout.

### d. Use Multiple Platforms

- Sign up for several platforms to maximize opportunities.
- Allocate specific time blocks for completing tasks on different platforms.

### e. Automate Where Possible

- Use browser extensions or apps to streamline workflows.
- Some tools can alert you to high-paying tasks or surveys.

## 6. Overcoming Challenges

### a. Dealing with Low Pay

- Combine surveys and microtasks with other income methods for better overall earnings.
- Focus on higher-value assignments as you gain experience.

## b. Avoiding Scams

- Be cautious of sites asking for upfront payments or excessive personal information.

- Stick to platforms with verified user reviews.

## c. Preventing Burnout

- Set daily or weekly goals to manage workload.

- Balance microtasks with other more engaging online ventures.

## 7. Scaling Beyond Surveys and Microtasks

While surveys and microtasks are great for supplemental income, they are not scalable for long-term financial stability. Consider transitioning to higher-paying opportunities as your time, skills, and resources grow. For instance:

- Use earnings from surveys to invest in online courses and develop new skills.

- Transition from microtasks to freelancing in areas like content writing, data analysis, or graphic design.

## CONCLUSION

Surveys and microtasks provide a simple and accessible way to earn money online with minimal investment of resources. They're especially useful for individuals seeking quick cash or as a stepping stone toward more substantial online income ventures. By using reputable platforms, working strategically, and avoiding scams, you can make the most of these opportunities and potentially open doors to more lucrative endeavours.

# CHAPTER 12

# Digital Marketing

## Exploring SEO, Social Media Ads, and More

Digital marketing is a powerful tool for reaching online audiences, promoting products or services, and building brands. Whether you're marketing your own business or offering services as a freelancer, mastering the essentials of digital marketing can be a lucrative skill set. This chapter explores the foundations of digital marketing, including SEO (Search Engine Optimization), social media advertising, content marketing, email campaigns, and more.

## 1. Understanding Digital Marketing

**Digital marketing** encompasses all marketing efforts using the internet and online technologies. It allows businesses to connect with their audience where they spend much of their time—online. The major channels include:

- **Search engines** (Google, Bing).
- **Social media platforms** (Facebook, Instagram, LinkedIn).
- **Email** and **websites**.

## 2. The Role of SEO in Digital Marketing

### a. What is SEO?

SEO, or Search Engine Optimization, is the practice of optimizing your website and content to rank higher on search engine results pages (SERPs). Better rankings lead to more organic traffic.

### b. Components of SEO

1. **On-Page SEO**
   - **Keywords:** Research and incorporate relevant keywords into your content.
   - **Content Quality:** Create engaging, valuable, and original content.
   - **Meta Tags:** Write compelling meta titles and descriptions.

2. **Off-Page SEO**

   - o **Backlinks:** Build links from reputable websites.

   - o **Social Signals:** Engage audiences on social media platforms to boost visibility.

3. **Technical SEO**

   - o **Site Speed:** Ensure your website loads quickly.

   - o **Mobile Optimization:** Create a responsive design for mobile users.

   - o **Secure Protocols:** Use HTTPS for security.

## 3. Social Media Advertising: Capturing Targeted Audiences

Social media platforms provide unparalleled opportunities to reach specific demographics with targeted ads.

### a. Key Platforms

1. **Facebook Ads**

   - o Advanced targeting options based on interests, behaviours, and demographics.

   - o Best for B2C businesses and community engagement.

2. **Instagram Ads**

   - o Visual-centric platform ideal for fashion, lifestyle, and creative niches.

   - o Leverages stories, reels, and carousel ads.

3.  **LinkedIn Ads**

   o   Focused on B2B marketing, offering professional targeting options.

   o   Great for recruiting, training, and industry-specific campaigns.


4.  **Google Ads (PPC)**

   o   Search Engine Marketing (SEM) ads that display on Google SERPs.

   o   Effective for driving traffic with high purchase intent.


## b. Crafting Effective Ads

1.  **Know Your Audience:** Use platform insights to identify the preferences and behaviours of your target demographic.

2.  **Compelling Creatives:** Use eye-catching visuals and engaging copy.

3.  **Clear Call-to-Actions (CTAs):** Guide users to take specific actions, like "Shop Now," "Learn More," or "Subscribe."


## 4. Content Marketing: Building Authority Through Valuable Information

### a. What is Content Marketing?

Content marketing involves creating valuable, relevant, and consistent content to attract and retain a clearly defined audience. The goal is to drive profitable customer actions.

### b. Types of Content

- **Blogs:** Write informative articles that answer audience questions.

- **Videos:** Tutorials, product demonstrations, and behind-the-scenes footage.

- **Infographics:** Present data visually for easy understanding.

- **E-books and Whitepapers:** Offer in-depth guides to establish authority.

### c. Strategy for Success

1. **Plan Your Content:** Develop a content calendar with consistent topics.

2. **Promote Widely:** Share content on social media, email, and collaborations.

3. **Track Metrics:** Use tools like Google Analytics to measure engagement.

## 5. Email Marketing: The Art of Staying in Touch

Email marketing remains one of the most cost-effective digital marketing strategies.

### a. Building an Email List

1. **Opt-In Forms:** Use website pop-ups or landing pages to collect emails.

2. **Lead Magnets:** Offer free resources (like e-books or templates) in exchange for email addresses.

### b. Creating Effective Campaigns

1. **Personalization:** Use the recipient's name and tailor the message to their interests.

2. **Segmentation:** Divide your audience into groups based on behaviour or preferences.

3.  **Clear CTA:** Encourage users to click, buy, or sign up.

### c. Tools for Email Marketing

- Mailchimp.
- ConvertKit.
- Constant Contact.

## 6. Analytics and Optimization: Improving Campaign Performance

### a. Why Analytics Matter

Analytics help measure the success of your digital marketing efforts. Without data, you can't determine what works or how to improve.

### b. Key Metrics

1.  **Traffic:** Track how many users visit your website or profile.
2.  **Engagement:** Monitor likes, shares, comments, and click-through rates (CTR).
3.  **Conversion Rates:** Measure how many visitors take desired actions, like purchasing or signing up.

### c. Tools for Analytics

- **Google Analytics:** Tracks website traffic and user behaviour.
- **Facebook Insights:** Measures social media engagement.
- **Ahrefs and SEMrush:** Analyse SEO performance.

## 7. Challenges in Digital Marketing

While digital marketing offers immense potential, it comes with challenges:

### a. Algorithm Changes:

Platforms like Google and Instagram frequently update their algorithms, affecting visibility.

### b. Competition:

High-quality content and ads are needed to stand out in crowded markets.

### c. Budget Constraints:

Ad campaigns require investment, which may be a hurdle for beginners.

## 8. Scaling Your Digital Marketing Efforts

### a. Automation

- Use tools like Hootsuite or Buffer to schedule and manage social media posts.
- Automate email campaigns with platforms like Mailchimp.

### b. Upskilling

- Take courses on platforms like Coursera, HubSpot, or Udemy to learn advanced strategies.

## c. Building a Team

- Hire specialists for content creation, SEO, and social media management.

## CONCLUSION

Digital marketing is one of the most versatile and scalable ways to earn money online. Whether you're running your own campaigns or providing marketing services to others, mastering SEO, social media ads, and content marketing can unlock immense potential. Focus on learning the fundamentals, staying adaptable, and continuously measuring performance to refine your strategies.

# CHAPTER 13

## Passive Income Strategies

### Building Income Streams That Run Themselves

Passive income is the holy grail for anyone seeking financial freedom and flexibility. Unlike traditional jobs or ventures that require constant effort, passive income involves creating systems that generate revenue with minimal ongoing work. Though establishing these streams requires upfront effort, the rewards are long-term and scalable.

This chapter explores various passive income strategies, offering insights into building revenue streams that can sustain and grow over time.

# 1. What Is Passive Income?

**Passive income** refers to earnings derived from activities or assets that require little to no active maintenance once they are established. Examples include royalties, dividends, rental income, and revenue from automated businesses.

### Key Characteristics of Passive Income:

- **Scalability:** Potential to grow without proportionate increases in effort.
- **Automation:** Operates independently with minimal intervention.
- **Diversity:** Offers multiple opportunities to mitigate risks.

# 2. Benefits of Passive Income

1. **Financial Freedom:** Enables you to focus on personal growth or other pursuits without constant financial worries.
2. **Flexibility:** Frees up time as income generation becomes less dependent on active work.
3. **Wealth Accumulation:** Provides ongoing revenue streams that can be reinvested to grow wealth.

# 3. Passive Income Strategies

### a. Content-Based Passive Income

1. **E-Books and Digital Products**
   - Write an e-book or create downloadable resources (templates, courses).

o Platforms: Amazon Kindle Direct Publishing (KDP), Gumroad, or Teachable.

**Example:** A recipe e-book for busy parents can generate consistent revenue.

## 2. YouTube Revenue

o Create and monetize a YouTube channel.

o Income sources include ads, sponsorships, and merchandise.

**Example:** A channel offering fitness tips can attract a loyal audience.

## 3. Online Courses

o Develop courses for platforms like Udemy, Skillshare, or Coursera.

o Use pre-recorded videos, downloadable PDFs, and quizzes to engage learners.

**Example:** A course on graphic design fundamentals can appeal to aspiring designers.

## b. Investment-Based Passive Income

## 1. Dividend Stocks

o Invest in stocks that pay regular dividends.

o Requires capital but offers long-term financial growth.

**Example:** Companies like Procter & Gamble and Coca-Cola are known for consistent dividends.

2. **Real Estate Investment**

- o Purchase rental properties or invest through Real Estate Investment Trusts (REITs).
- o Generates income through rent or property appreciation.

**Example:** Renting out a vacation property during peak seasons can yield high returns.

3. **Peer-to-Peer Lending**

- o Lend money via platforms like LendingClub or Prosper.
- o Earn interest on loans.

**Example:** Funding small business loans can yield steady returns.

## c. Online Business Models

1. **Print-on-Demand Services**

- o Sell customized merchandise like T-shirts or mugs via platforms like Printful or Teespring.
- o Designs can be uploaded once and sold repeatedly.

**Example:** A graphic designer creates humorous quotes for coffee mugs and sells them online.

2. **Dropshipping**

- o Operate an e-commerce store without managing inventory.
- o Platforms: Shopify, Oberlo.

**Example:** Selling tech gadgets via an online store while suppliers handle fulfilment.

3. **Membership Sites or Subscriptions**

  - Create a subscription-based platform offering exclusive content.

  - Platforms: Patreon, Substack.

**Example:** A personal finance coach offers monthly newsletters and webinars for a fee.

## 4. Steps to Build a Passive Income Stream

### Step 1: Identify Your Skills and Interests

- Choose a strategy that aligns with your strengths or passion.

### Step 2: Research and Plan

- Evaluate market demand, competition, and potential revenue.

- Draft a business plan outlining steps, costs, and goals.

### Step 3: Invest Time and Resources

- Dedicate time to creating high-quality content, products, or investments.

### Step 4: Automate Where Possible

- Use tools and platforms to handle repetitive tasks.

- Example: Automating email marketing or customer support.

### Step 5: Diversify

- Establish multiple income streams to reduce reliance on any one source.

### Step 6: Monitor and Optimize

- Track performance and tweak strategies to improve results.

## 5. Common Challenges and How to Overcome Them

### Challenge 1: High Initial Effort

- **Solution:** Break projects into manageable steps and set milestones.

### Challenge 2: Risk of Failure

- **Solution:** Start small to test ideas before scaling.

### Challenge 3: Maintaining Relevance

- **Solution:** Stay updated on market trends and consumer preferences.

## 6. Inspiring Success Stories

### Example 1: Pat Flynn (Smart Passive Income)

- Transitioned from a layoff to building a multi-stream passive income business.
- Generates revenue from e-books, courses, and affiliate marketing.

### Example 2: Sarah Titus

- Created a successful printable business, earning six figures monthly.

**Example 3: Graham Stephan**

- A real estate investor who scaled his income through YouTube and online courses.

## 7. Scaling Passive Income Over Time

### a. Reinvestment

- Use earnings to fund additional streams or upgrade existing systems.

### b. Outsourcing

- Delegate tasks like customer support, content creation, or technical maintenance.

### c. Expanding Platforms

- Increase visibility by leveraging multiple platforms for your content or products.

## CONCLUSION

Passive income is not just about earning money; it's about designing a life of financial independence and security. With careful planning, consistent effort, and a willingness to learn, you can create income streams that sustain themselves while giving you the freedom to pursue your passions. From content creation to smart investments, the possibilities are endless—your journey to financial autonomy begins here.

# CHAPTER 14

## Avoiding Scams

### Staying Safe in the Online Marketplace

As online opportunities grow, so do the risks of scams and fraudulent schemes. The anonymity of the internet can make it a fertile ground for unethical actors preying on unsuspecting individuals. To navigate the online marketplace safely, understanding common scams, red flags, and protective strategies is essential.

This chapter provides a comprehensive guide to recognizing and avoiding scams, ensuring that your efforts to make money online are secure and rewarding.

# 1. The Importance of Vigilance

The internet has revolutionized earning opportunities, but it has also created a digital minefield. Fraudulent schemes range from phishing emails and fake investment opportunities to job scams and identity theft.

- **Statistics:** Cybercrime caused global losses of over $6 trillion annually, a figure projected to rise.
- **Target Groups:** Scammers often target students, housewives, and beginners seeking flexible income options.

# 2. Common Online Scams

## a. Phishing Scams

- Scammers impersonate trusted entities (banks, employers, or platforms) to steal personal information.
- **Example:** Emails with fake links asking for login credentials or payment details.

## b. Fake Job Offers

- Scams promising high-paying remote jobs with upfront fees for "training" or "equipment."
- **Example:** A job offer requiring a "registration fee" for a work-from-home position.

## c. Investment Scams

- Fraudulent schemes claiming guaranteed high returns with no risk.

- **Example:** Ponzi schemes and fake cryptocurrency trading platforms.

### d. Marketplace Scams

- Selling counterfeit products or never delivering purchased goods.
- **Example:** An e-commerce platform offering luxury items at too-good-to-be-true prices.

### e. Pyramid Schemes

- Recruitment-based schemes disguised as business opportunities.
- **Example:** Promises of earning by recruiting others rather than selling actual products or services.

### f. Identity Theft

- Collecting personal data through fake forms, contests, or surveys.
- **Example:** A website requiring excessive personal information for a simple signup.

## 3. Recognizing Red Flags

To avoid falling victim to scams, learn to identify warning signs.

1. **Unrealistic Promises**
   - Claims of instant wealth or guaranteed earnings with minimal effort.
   - **Rule of Thumb:** If it sounds too good to be true, it probably is.

2. **Pressure Tactics**

   o Demands for immediate action, such as "limited-time offers" or "urgent payment."

3. **Unverified Platforms**

   o Lack of verifiable credentials, reviews, or testimonials.

4. **Upfront Fees**

   o Requests for payment before delivering services or opportunities.

5. **Unsecured Websites**

   o Absence of HTTPS (a secure connection) or a professional website design.

6. **Lack of Transparency**

   o Vague information about the company, opportunity, or terms of payment.

## 4. Tips for Staying Safe Online

### a. Verify Opportunities

- Research companies, platforms, or individuals offering opportunities.
- Use trusted review sites like Glassdoor, Trustpilot, or Reddit.

### b. Protect Personal Information

- Avoid sharing sensitive data, such as Social Security numbers or bank details, unless verified as necessary.

### c. Use Secure Payment Methods

- Opt for methods offering buyer protection, such as PayPal or credit cards.

- Avoid wire transfers and cryptocurrency payments for unverified transactions.

### d. Enable Two-Factor Authentication (2FA)

- Use 2FA on all accounts to enhance security against unauthorized access.

### e. Check for Contact Information

- Legitimate companies provide detailed contact information, including a physical address and customer service channels.

### f. Stay Updated

- Follow cybersecurity updates to learn about new scam tactics.

## 5. Tools and Resources for Protection

### 1. Antivirus and Anti-Malware Software

- Tools like Norton, McAfee, or Bitdefender safeguard your devices.

### 2. Fraud Reporting Agencies

- Report scams to platforms like FTC (Federal Trade Commission) or Action Fraud.

3. **Password Managers**

   - Use tools like LastPass or Dashlane for creating and storing strong passwords.

4. **Browser Extensions**

   - Ad-blockers and anti-phishing extensions (like HTTPS Everywhere) enhance browsing safety.

5. **Online Scam Checkers**

   - Websites like ScamAdviser analyse the legitimacy of online platforms.

# 6. What to Do If You're Scammed

Despite precautions, scams can still happen. Here's how to respond:

## a. Stop Communication

- Cease all contact with the suspected scammer immediately.

## b. Freeze Accounts

- Contact your bank or financial institution to halt transactions.

## c. File a Report

- Notify local authorities and relevant cybersecurity organizations.

## d. Warn Others

- Share your experience on review platforms or forums to alert others.

## e. Monitor for Identity Theft

- Keep track of your credit report and set fraud alerts on your accounts.

# 7. Real-Life Scam Examples and Lessons

## Case Study 1: The Freelance Fake Client

- **Scenario:** A freelancer is approached with a lucrative project but asked to pay a "security deposit."
- **Outcome:** The client disappears after receiving payment.
- **Lesson:** Never pay to secure freelance projects.

## Case Study 2: The Social Media Influencer Trap

- **Scenario:** An aspiring influencer is offered a sponsorship deal requiring them to buy products first.
- **Outcome:** The sponsor is a fake, and the money is lost.
- **Lesson:** Genuine sponsorships do not require upfront purchases.

# 8. Cultivating a Safety-First Mindset

## a. Question Everything

- Approach all online opportunities with a healthy degree of skepticism.

## b. Educate Yourself

- Learn about common scams and teach friends and family about them.

## c. Trust Your Instincts

- If something feels off, investigate further or avoid it altogether.

## d. Build a Network

- Join communities of like-minded individuals who can provide guidance and warnings about potential scams.

## CONCLUSION

The internet is full of opportunities, but staying safe requires vigilance, knowledge, and preparation. By recognizing red flags, using protective tools, and cultivating a cautious mindset, you can safeguard yourself while exploring legitimate ways to make money online. Remember, success in the online marketplace isn't just about finding opportunities—it's about avoiding pitfalls that could derail your progress.

# CHAPTER 15

## Expanding Horizons

## Exploring Additional Legitimate Online Money-Making Avenues

The world of online income is vast, continually evolving, and brimming with opportunities beyond the well-known methods discussed in earlier chapters. For those willing to explore further, there are other equally authentic and rewarding ways to earn money online. This chapter delves into these alternative methods, providing insights into their potential and strategies for success.

## 1. Online Market Research Participation

### Overview

Companies and researchers rely on public input to refine products, services, and strategies. Participating in market research studies, focus groups, and usability testing provides an easy way to earn money while contributing to innovation.

### How It Works

- Register on platforms like Respondent, UserTesting, or Survey Junkie.
- Get paid for tasks like reviewing websites, testing products, or participating in interviews.

### Potential Earnings

- Usability tests: $10–$60 per session.
- Focus groups or interviews: $50–$250 per hour.

**Tips for Success**

- Complete your profiles thoroughly to match with relevant opportunities.
- Respond promptly to invitations for high-demand studies.

## 2. Selling Stock Photos, Videos, and Music

**Overview**

If you have a knack for photography, videography, or music production, licensing your creations to stock platforms is a lucrative option. Content creators, marketers, and businesses constantly seek high-quality media.

**How It Works**

- Submit your work to platforms like Shutterstock, Adobe Stock, or Pond5.
- Earn royalties whenever someone downloads your content.

**Potential Earnings**

- Photos: $0.25–$5 per download.
- Videos or music: $20–$200 per license.

**Tips for Success**

- Focus on trending or evergreen content that meets market demands.
- Use descriptive and keyword-rich metadata for better visibility.

## 3. Participating in Domain Name Flipping

### Overview

Domain name flipping involves buying undervalued domain names and reselling them for a profit. This niche market thrives on creativity and foresight.

### How It Works

- Purchase domain names through registrars like GoDaddy or Namecheap.
- List them on marketplaces such as Sedo or Flippa.

### Potential Earnings

- Profit margins vary significantly; some domains sell for hundreds or even thousands of dollars.

### Tips for Success

- Focus on domains with high commercial potential or trending keywords.
- Avoid trademarked terms to prevent legal issues.

## 4. Creating and Selling NFTs

### Overview

Non-fungible tokens (NFTs) offer a revolutionary way to monetize digital assets like art, music, or collectibles. With blockchain technology, creators can earn from initial sales and future resales.

### How It Works

- Create digital assets and mint them as NFTs on platforms like OpenSea or Rarible.

- Market your NFTs to potential buyers through social media or NFT communities.

### Potential Earnings

- Earnings depend on demand, rarity, and your marketing efforts.

### Tips for Success

- Focus on creating unique, high-quality digital assets.

- Engage with the NFT community to build credibility and interest.

## 5. Crowdsourcing and Open Innovation Platforms

### Overview

Crowdsourcing platforms allow individuals to contribute ideas, solutions, or creative work to businesses and organizations.

### How It Works

- Sign up on platforms like Innocentive, 99designs, or Tongal.

- Participate in contests or projects tailored to your skills.

### Potential Earnings

- Rewards range from $50 to $100,000, depending on the project's complexity and sponsor.

### Tips for Success

- Choose challenges aligned with your expertise.
- Review past winning entries to understand what sponsors value.

## 6. Transcription Services

### Overview

Transcribing audio or video files into written text is a straightforward way to earn online, ideal for those with excellent listening and typing skills.

### How It Works

- Join platforms like Rev, TranscribeMe, or GoTranscript.
- Complete transcription tasks and earn based on file length and difficulty.

### Potential Earnings

- $15–$25 per audio hour for beginners.

### Tips for Success

- Invest in transcription software to improve speed and accuracy.
- Choose files with clear audio to avoid frustration.

## 7. Participating in Virtual Events

### Overview

As virtual events become mainstream, opportunities to earn as a host, moderator, or event assistant have increased.

### How It Works

- Offer services on platforms like Zoom or Eventbrite.
- Handle tasks such as managing event schedules, moderating Q&A sessions, or troubleshooting technical issues.

### Potential Earnings

- $20–$50 per hour, depending on event size and complexity.

### Tips for Success

- Develop technical skills for virtual event platforms.
- Build a professional profile highlighting your organizational expertise.

## 8. Writing Scripts for Videos or Ads

### Overview

Scriptwriting for YouTubers, podcasters, or businesses is a niche skill in high demand.

### How It Works

- Connect with clients on platforms like Upwork or Fiverr.
- Write engaging and compelling scripts tailored to the client's audience and goals.

### Potential Earnings

- $50–$500 per script, depending on length and complexity.

### Tips for Success

- Study successful scripts in your target niche.
- Use storytelling techniques to make content engaging.

## 9. Online Language Translation

### Overview

Fluency in multiple languages can open doors to online translation work.

### How It Works

- Sign up on platforms like Gengo, Unbabel, or ProZ.
- Translate documents, websites, or subtitles.

### Potential Earnings

- $0.03–$0.10 per word.

### Tips for Success

- Focus on industries like legal, medical, or technical translation for higher rates.
- Maintain accuracy and cultural sensitivity.

## 10. Testing Apps and Games

### Overview

App and game developers require user feedback to refine their products. Paid testing opportunities provide compensation for your insights.

### How It Works

- Join platforms like TesterWork or PlaytestCloud.
- Provide feedback on usability, functionality, and design.

### Potential Earnings

- $10–$30 per test, depending on complexity.

### Tips for Success

- Be thorough and articulate in your feedback.
- Use a range of devices to increase eligibility.

# CONCLUSION

The digital world offers endless ways to earn a legitimate income. Whether you're leveraging your skills, creativity, or interests, these alternative methods can complement or even surpass traditional avenues. The key lies in exploring these opportunities with curiosity and dedication, unlocking new horizons for sustained success in the online economy.

# CHAPTER 16

## Creating Your Sustainable Path

## From Part-Time to Full-Time Online Success

The transition from part-time to full-time online work is a transformative journey that requires strategic planning, perseverance, and adaptability. While earning online can start as a side hustle, achieving long-term sustainability demands that you balance effort and growth. In this chapter, we explore how to transition your online endeavours into a reliable, full-time income source.

## 1. The Importance of a Sustainable Online Strategy

Sustainability is the backbone of long-term success in the online marketplace. Whether you are freelancing, selling products, or managing an affiliate business, establishing consistent and scalable income streams is essential.

- **Why It Matters:** A sustainable strategy ensures financial stability, protects against burnout, and provides room for growth.
- **Mindset Shift:** Treating your online work as a professional enterprise rather than a casual pursuit is key to sustainability.

## 2. Defining Your Vision of Full-Time Success

### a. Set Clear Goals

- Define what "full-time success" looks like for you—financially, professionally, and personally.

- **Example Goals:**
    - o Earn $5,000 monthly through diversified income streams.
    - o Work no more than 40 hours per week while maintaining flexibility.

## b. Calculate Your Financial Needs

- Assess your monthly expenses to determine how much you need to sustain your lifestyle.
- Include costs like taxes, health insurance, and savings in your calculations.

## c. Plan for Growth

- Consider how to scale your efforts over time to achieve financial freedom.

# 3. Diversifying Income Streams

## Why Diversification Matters

Relying on a single income source is risky in the volatile online marketplace. Diversification offers stability and protects against downturns in specific areas.

## Examples of Diversified Online Income Streams

1. Freelancing projects (writing, graphic design, virtual assistance).
2. Affiliate marketing partnerships.
3. Selling digital products like e-books or courses.

4.  Running a blog with ad revenue and sponsorships.

5.  Offering consulting or coaching services.

**How to Manage Multiple Streams**

- Allocate time for each stream based on its growth potential.

- Use tools to automate repetitive tasks, like email marketing or social media scheduling.

## 4. Scaling Your Online Business

### a. Expanding Your Client Base

- Build a reputation in your niche through excellent service, testimonials, and networking.

- Platforms like LinkedIn and Fiverr can help you reach more clients.

### b. Building a Personal Brand

- Create a professional website or portfolio showcasing your skills and successes.

- Engage with audiences on social media to position yourself as an authority in your field.

### c. Leveraging Outsourcing

- Delegate tasks to focus on high-value activities.

- Examples of tasks to outsource: administrative duties, content creation, or website management.

## d. Scaling Digital Products

- Add new products or services to your offerings, such as advanced courses or exclusive content.

- Use feedback from your audience to refine and expand your offerings.

## 5. Establishing a Work-Life Balance

## Challenges of Full-Time Online Work

- Blurred boundaries between personal and professional life.
- Potential for overworking due to lack of fixed hours.

## Strategies for Balance

1. **Set a Schedule:**

   o Establish regular working hours and stick to them.

2. **Create a Dedicated Workspace:**

   o Separate your work area from personal spaces to enhance focus.

3. **Take Breaks:**

   o Schedule regular intervals to rest and recharge.

4. **Prioritize Self-Care:**

   o Invest time in hobbies, exercise, and social activities.

# 6. Adapting to Market Changes

## Staying Relevant in a Dynamic Market

- The online marketplace evolves rapidly. To remain successful, adapt to new trends, technologies, and audience needs.

## Ways to Stay Ahead

1. **Continuing Education:**
   - Enroll in courses or attend webinars to upskill.
2. **Networking:**
   - Join online communities and forums to exchange knowledge and opportunities.
3. **Experimentation:**
   - Test new ideas, platforms, or strategies to find what works best.

# 7. Challenges and How to Overcome Them

## Challenge 1: Income Inconsistency

- **Solution:** Build a financial buffer to handle low-income months.

## Challenge 2: Competition

- **Solution:** Differentiate yourself by offering unique value and quality services.

**Challenge 3: Staying Motivated**

- **Solution:** Celebrate milestones and maintain a vision board of your goals.

## 8. Real-Life Stories of Online Success

### Case Study 1: The Transition of a Blogger

- Started as a part-time food blogger.

- Grew income through ad revenue, sponsorships, and a best-selling recipe e-book.

- Achieved full-time income within two years.

### Case Study 2: From Freelancer to Agency Owner

- Began freelancing in web development.

- Outsourced tasks and scaled services into a full-fledged digital agency.

- Revenue tripled after expanding the team and offerings.

## 9. Building a Long-Term Vision

### Think Beyond Earnings

- Focus on creating a legacy, such as a recognized brand or a business that can operate without you.

### Plan for Retirement

- Save and invest a portion of your online earnings to secure your future.

**Give Back to the Community**

- Mentor aspiring online entrepreneurs or contribute to causes that align with your values.


## CONCLUSION


Transitioning from part-time to full-time online work is both challenging and rewarding. It requires careful planning, continuous learning, and an entrepreneurial mindset. By diversifying income, scaling operations, and staying adaptable, you can build a sustainable career in the online world. Your path to success is as unique as your talents and ambitions—embrace the journey with determination and vision.

With this chapter, our journey through legitimate and authentic ways to make money online comes to an end, but your journey is just beginning. Stay persistent, stay safe, and continue building the life you desire through online success.

**CONCLUSION CHAPTER**

**Crafting Your Digital Success**

**A Comprehensive Guide to Thriving Online**

The journey toward making a legitimate income online is filled with possibilities, but success hinges on your approach, consistency, and adaptability. As a brilliant business consultant, Zaheer Ahmed Shaik emphasizes that online success isn't about shortcuts but about cultivating a strong foundation, strategic mindset, and unwavering dedication. Here's a concise summary of all the important insights and best advice from the previous chapters to guide you on your path.

## 1. Begin with Clarity and Planning

- **Define Goals:** Understand your financial needs and long-term aspirations.
- **Select Your Niche:** Focus on areas where your skills, interests, and market demand align.
- **Learn Continuously:** Stay updated with trends and invest in personal development.

## 2. Explore Diverse Opportunities

Zaheer Ahmed Shaik highlights the importance of diversifying your income streams to minimize risk and enhance stability. Here are the main avenues covered:

- **Freelancing:** Offer skills like writing, graphic design, or virtual assistance.

- **Blogging and Content Creation:** Build a platform for ad revenue, sponsorships, or product sales.

- **Affiliate Marketing:** Promote products authentically and earn through commissions.

- **E-commerce and Dropshipping:** Sell physical or digital products with minimal inventory.

- **Online Tutoring:** Share expertise in academic or niche fields.

- **Digital Marketing:** Leverage SEO, social media ads, and analytics to grow businesses.

## 3. Focus on Scalability

Building sustainable income streams requires scalability:

- Automate tasks where possible (e.g., email campaigns, social media scheduling).

- Delegate work to freelancers or team members as your workload increases.

- Invest in long-term opportunities like creating digital products or establishing passive income streams.

## 4. Embrace Adaptability

Online markets change rapidly. To thrive:

- **Stay Ahead of Trends:** Engage with industry news and adopt emerging technologies.

- **Experiment:** Test new platforms, strategies, or content styles.

- **Pivot When Needed:** Don't be afraid to shift focus if an approach is no longer effective.

## 5. Prioritize Trust and Credibility

Success online depends on your reputation.

- Always deliver quality, whether it's a product, service, or content.
- Avoid scams and unethical practices that could damage your credibility.
- Engage with your audience authentically to build trust and loyalty.

## 6. Time Management and Work-Life Balance

Zaheer emphasizes the importance of maintaining balance:

- Create a dedicated workspace to separate personal and professional life.
- Set working hours and honour breaks to prevent burnout.
- Schedule downtime to recharge and pursue other passions.

## 7. Overcome Challenges with Resilience

Every journey has its hurdles, but they're stepping stones to success.

- **Income Fluctuations:** Build a financial buffer for lean periods.
- **High Competition:** Differentiate yourself with unique skills and exceptional service.
- **Motivation Slumps:** Revisit your goals and celebrate milestones to stay inspired.

# 8. Tips for Long-Term Success

- **Think Big:** Consider how your current efforts can evolve into a larger enterprise.

- **Network Wisely:** Connect with other professionals to exchange ideas and opportunities.

- **Invest in Yourself:** The best investment is in your knowledge and skills.

### Final Thoughts from Zaheer Ahmed Shaik

"Online success is not a race; it's a marathon. Focus on consistency, integrity, and value creation. The internet is an incredible equalizer, offering opportunities to anyone willing to put in the effort. Treat your online ventures as a business, keep learning, and stay adaptable. The path to financial freedom and fulfilment lies in your ability to embrace challenges, seize opportunities, and remain committed to your vision."

With these strategies, tips, and insights, you are equipped to craft a thriving, sustainable digital career—one that aligns with your aspirations and brings lasting success.

# REFERENCES

The following resources, books, articles, and platforms provided valuable insights and information that contributed to the creation of *The Ultimate Guide to Making Money Online: Legitimate and Authentic Opportunities for Students, Women, Housewives, and Remote Workers*. Each reference has been carefully selected to ensure credibility and relevance to the topics discussed.



1. **Freelancing Platforms and Marketplaces**

   - Websites: Upwork, Fiverr, Toptal, Freelancer.com

   - Articles on building freelancing careers from platforms like HubSpot and Forbes


2. **Blogging and Content Platforms**

   - Resources: WordPress.org, Medium.com, Wix

   - Books: *ProBlogger: Secrets for Blogging Your Way to a Six-Figure Income* by Darren Rowse and Chris Garrett


3. **Affiliate Marketing**

   - Guides: Amazon Associates Program, ClickBank, CJ Affiliate Resources

   - Books: *Affiliate Marketing For Beginners: Build Your Passive Income and Start Your Freedom Journey* by Michael Ezeanaka


4. **E-Commerce and Dropshipping**

   - Platforms: Shopify, WooCommerce, Oberlo, AliExpress

   - Articles from Shopify Blog and Oberlo's guide to dropshipping

5. **Online Tutoring**

   o   Websites: Tutor.com, Chegg Tutors, VIPKid, Udemy, and Teachable

   o   Articles on creating online courses from Thinkific and Coursera blogs

6. **Digital Marketing and SEO**

   o   Resources: Google Digital Garage, Moz Blog, Neil Patel's SEO Guides

   o   Books: *SEO 2023: Learn Search Engine Optimization with Smart Internet Marketing Strategies* by Adam Clarke

7. **Content Creation**

   o   Platforms: YouTube Creator Academy, Buzzsprout (for podcasting), and Canva

   o   Books: *YouTube Secrets: The Ultimate Guide to Growing Your Following and Making Money as a Video Influencer* by Sean Cannell and Benji Travis

8. **Graphic Design and Creative Ventures**

   o   Platforms: Adobe Creative Cloud, Canva, Dribbble, and Behance

   o   Guides: Articles from Creative Bloq and Smashing Magazine

9. **Surveys and Microtasks**

   o   Websites: Amazon Mechanical Turk, Swagbucks, InboxDollars, Clickworker

10. **Passive Income Strategies**

   o   Books: *Passive Income, Aggressive Retirement* by Rachel Richards

   o   Resources from Smart Passive Income by Pat Flynn

11. **Avoiding Online Scams**

   o Guides from: FTC Consumer Information, BBB (Better Business Bureau), and Scamwatch


12. **Virtual Assistance**

   o Platforms: Zirtual, Time Etc., and Boldly

   o Articles on becoming a VA from Virtual Savvy Blog


13. **General Financial Advice and Planning**

   o Books: *Rich Dad Poor Dad* by Robert Kiyosaki

   o Articles from Investopedia and Entrepreneur


14. **Online Income Research and Case Studies**

   o Reports and studies from Statista, Pew Research Center, and GlobalWebIndex

   o Whitepapers and surveys on digital workforce trends


15. **Author's Experience and Expertise**

   o Zaheer Ahmed Shaik's professional knowledge in business consulting, digital marketing, and web development contributed significantly to the authenticity and practicality of this book.


This references list ensures that anyone can have access to credible sources for further exploration. The integration of diverse platforms and publications aims to provide a comprehensive understanding of legitimate online income opportunities.